SMALL FURRY ANIMALS

Harvest Mouse

SMALL FURRY ANIMALS

Harvest Mouse

Ting Morris

Illustrated by Graham Rosewarne

W

FRANKLIN WATTS

LONDON • SYDNEY

You are standing beside a field full of poppies. As you watch, the flowers and grasses sway in the wind.

The longer you look, the more you see. A tiny creature, no bigger than your finger, is swinging between two tall stems of grass. You see orange fur, a white belly, and round, black eyes. Who is this furry acrobat?

Turn the page and take a closer look.

You've spotted a harvest mouse. This tiny creature is the smallest member of the mouse family. It's an acrobatic climber, but this one is clinging to the stem like a statue. That's because she's spotted you too. She probably heard you nearby. When it's safe, she'll scamper away under the cover of the poppies.

6

MINI MAMMAL

Harvest mice are mammals. A mammal has hair or fur on its body to keep it warm. Baby mammals are fed milk from their mother's body. Human beings are mammals too.

Nimble climbers

You can identify a harvest mouse by its orange-coloured coat, small ears and stub nose. Weighing about 6 grams and measuring just over 5 centimetres long, it uses its tail as an extra hand for gripping. Harvest mice often hang from their hind feet and tail, and use their front paws to pick and hold food.

TAIL POINTS

Length: 5–7.5 centimetres, about as long as the mouse's body.

Functions: extra limb – used as a balancing tool when climbing up, and as a brake when climbing down.

Tip: prehensile – the last 2.5-centimetre portion is used to grasp and wrap around objects for support.

Mice

Mice belong to a group of mammals called rodents, which means 'gnawing'. That's because they have strong front teeth that they use to gnaw their food. Mice, rats and their relatives are the most widespread of all rodents. They are usually seen as pests, because many carry diseases and live close to us, eating our food.

The harvest mouse is on night duty. She has left her flimsy nest at ground level and moved up the grass stem. Here she is building a big, strong nest where her babies will be safe.

Watch how skilfully she builds. First she slits the stems with her front teeth so they bend. She pulls them into position to make a framework. Then she shreds all the green leaves around the stems, without cutting any off, and weaves them into a ball.

8

Breeding nests

Female harvest mice build breeding nests for their young, and the fathers don't help. The builder weaves a round nest well off the ground, 35 to 60 centimetres up. Once the outer framework is complete, she lines the nest from the inside. She collects thistledown and chews up leaves and flower heads for stuffing. The nest is finished when the inside is completely filled with soft materials.

GRASSY HOME

Harvest mice are at home wherever tall grasses and long-stemmed plants grow. You'll find them in reed beds, ditches, wild meadows and even overgrown gardens. In the winter, they may be found in warmer places, such as hollow logs, empty birds' nests or in haystacks.

9

The harvest mouse is in no rush. She's built many nests like this before, and her babies won't be born for a while yet. Now she's finishing the nest from the inside. She pulls leaf-ends through the walls, chews them up, and fills gaps with the mush. She is a builder, weaver and plasterer all in one.

But what's the mouse up to now? She's bringing in more lining, including a feather she has found. Will she be able to pack in any more?

10

Mating

Harvest mice mate when they are five to six weeks old. Most litters are born in August and September.

SINGLE NURSERIES

If the male doesn't leave after mating, the female chases him away. Harvest mice are pregnant for about 18 days and, during that time, they become twice as heavy. The mother-to-be builds a new nursery nest for each litter.

Fast breeders

All mice breed at almost any time of the year and have many young. They have lots of enemies and are on the menu of most meat-eating animals, so many mice don't survive long. Harvest mice have three litters a year.

The pink ball in the middle of the nest is six baby mice huddled together for warmth. Each one is just a little bigger than your fingernail and weighs hardly anything.

The young were born last night. They can't move much and can't see yet, but they can squeak! After every feeding, the mother mouse washes her babies and cleans out the nursery. The newborns will grow fast and, in a couple of days, they'll have their first fur.

SQUEAK, SQUEAK!

Baby harvest mice don't cry, as a noisy nest might attract hungry enemies. But on the first day they squeak if touched. If a mother has to move her young from one nest to another and drops a baby, it will cry out. She follows its squeaks and searches the grass until she finds it.

The newborns

Harvest mice are usually born at night or early in the morning. They are a little more than 15 millimetres long, with a tail about 9 millimetres long. Within a day, each baby's pink skin changes to grey. By the third day, light brown hair begins to grow. By the fifth day, the babies begin to crawl around the nest.

On nest duty

The mother makes sure she is not seen when she slips into the nest to feed her young. She lies above the babies so they can all suck milk from her nipples. She washes them with her tongue and eats anything that gives off smells, to keep hungry prowlers away from the nest.

13

The babies are nine days old now, and they can see and hear. They still drink their mother's milk, but today they are tasting their first solid food. Their mother has brought back seeds and grains, and she chews everything up for them first.

The little ones groom themselves now, but they've still got a lot to learn. One has just toppled over. It's a good thing the nest can stretch with the growing brood.

GROWTH CHART

*6 days: Furred back.
Dull-brown or sandy coat.
8 days: Eyes and ears open.
Front teeth come through.
9 days: Eat first solid food.
Move about in the nest.
11–12 days: Explore outside the
nest, but come back to sleep.
15–16 days: Mother stops
suckling and leaves the litter.
Youngsters are on their own.*

BABY DEFENCE

*Young harvest mice are born with
an instinct for self-defence. If the
nest is disturbed, the young stretch
out together. They do this a number of
times so passing enemies leave the nest
alone, fearing that a big animal is
inside. As soon as young harvest
mice have teeth, they show
them if they sense
danger.*

By the scruff of the neck

When carrying a baby, the mother mouse holds it in her teeth by the scruff of its neck. As she grips the skin, the young mouse goes stiff. It clenches its legs and feet, and pulls its tail up under its belly. This makes it easy to transport.

When the youngsters play outside the nest, they stay close to home and hide if they hear a rustling sound. The nest is a bit battered now, but the young mice still sleep in it. Their mother won't be coming back. She'll soon be having another litter and is busy making a new nest.

How many harvest mice can you see? There are just four left. Two others fell from the nest, and a hungry toad swallowed them up. It's getting dangerous now. A crow has his eye on the nest. It's time for the mice to move on.

hawk

Big enemies

Harvest mice can live for up to 18 months, but most live for less than six weeks. These bite-sized mice make an easy meal for many other animals. They are hunted by weasels, foxes, cats, toads and many birds, including hawks and owls. Near water, herons pick them off reeds. Even larger mice and rats eat their tiny relatives.

Nestlings in danger!

Young harvest mice are at the greatest risk, and many are caught on their first outing. Nestlings aren't safe either. Weasels don't mind a steep climb to check what's in the nest and on the menu.

owl

fox

cat

toad

heron

weasel

17

Now two harvest mice are quarrelling over food. The smaller one is a youngster, and he's come to the wrong place for his grain. The older mouse is striking out and chattering angrily. It sounds like 'zt - ick - zt - tick - zt - ick.' Perhaps he's saying, 'Get off my patch!'

MOUSE QUARREL

Harvest mice don't make friends. To mark their territory, mice rub their bottom against the underside of grass leaves and stems so their urine and excrement stick to the plants. If another mouse doesn't get the smelly message, they fight it out. Quarrelling harvest mice sit on their back legs and tail. They make chattering noises and box with their front paws.

Seed pickers

In the summer, harvest mice eat green plant sprouts, but their favourite meal is grain. They pick the seeds off the ground or harvest them from plants by bending the stem to the ground, where they can bite off the seeds. For safety reasons, dinner is usually eaten high off the ground. The mouse holds the seed in its front paws and gnaws it. Harvest mice don't damage crops much, and can even help farmers by keeping down insect pests.

MOUSE MENU

CEREAL GRAINS
oats, wheat, millet

WILD PLANTS
reed seeds, common reeds,
hogweed seeds,
bush grass, couch grass,
bulrushes, cowslip flowers,
rice, maize

FRUIT
blackberries, raspberries,
hips, haws,
hedgerow fruit mix

MEAT
flies, midges,
grasshoppers, bush crickets,
caterpillars, ladybirds,
moths, butterflies,
beetles,
small birds' eggs

DRINKS
nectar, water,
honeydew from aphids

19

This morning a cool autumn sun was shining on the cornfield when the harvest mice heard a loud rattling noise. They stayed in their nests or hid in the undergrowth. By afternoon, there was only stubble left. The mice were homeless, but it takes more than a combine harvester to beat a clever harvest mouse.

A few trespassers have stowed away in stacks of corn. They are about to make a journey to the barn down the road. Others are scurrying through the stubble to nearby meadows and the reed bed. Will there be enough time for the females to build other breeding nests?

Hedge cutters and harvesting machines

From one day to another, a harvest mouse's territory can change or disappear completely. An overgrown meadow may be mown. Hedgerows are trimmed or uprooted.
In the autumn, combine harvesters move into cornfields and cut down all the grain.

This young mouse escaped to the reed bed. It's October and she is soon going to have her own family, but the nest building is not going well. She found a good place above the water, but things kept going wrong. Once she fell into the water and had to swim to safety. This is her third nest, and now she has to stop again because it is raining. She'll have to look for shelter now.

FIRST TRY

Young mice don't receive nest-building lessons and often make two or three attempts before they get it right. It's difficult to pull all the loose leaves into a round ball. In bad weather, the building work can take more than a week. Experienced mothers can build a nest in two or three hours.

Keeping warm

Harvest mice stay in their nest when it's cold and wet. The nest is somewhat waterproof and, inside, the mouse curls up into a ball to keep warm. If caught in the rain, a harvest mouse shelters in the thick undergrowth or in an empty breeding nest. If a harvest mouse gets soaked, it can die in less than an hour.

UP THE NEXT STALK

During the summer, many harvest mice live in a single field. Male mice don't worry much about territories, but females keep other females out of their own breeding range. In thick grass, there is about 3 metres between nests. In short grass, there is about 6 metres between nests.

Where in the world?

In France and Italy, harvest mice live in cornfields. In Russia and Asia, they often nest in rice plants. In Japan, the little mice sometimes settle among strawberry crops. Harvest mice don't mind water and can swim. They live in grassy sand dunes, salt marshes and the banks of rivers, streams and ditches, where they build nests in reeds above the water. The American plains harvest mouse lives on the open prairie.

23

These three youngsters have shared this nest all through the winter. We saw their mother building it in the reeds when the weather turned wet and cold. When her babies were big enough to look after themselves, she quickly left and moved to a dry meadow. Now it's almost spring again, and the three youngsters will soon say goodbye. Perhaps one of them will find the poppy field. They'll be pleased if they do, as wheat is growing there now.

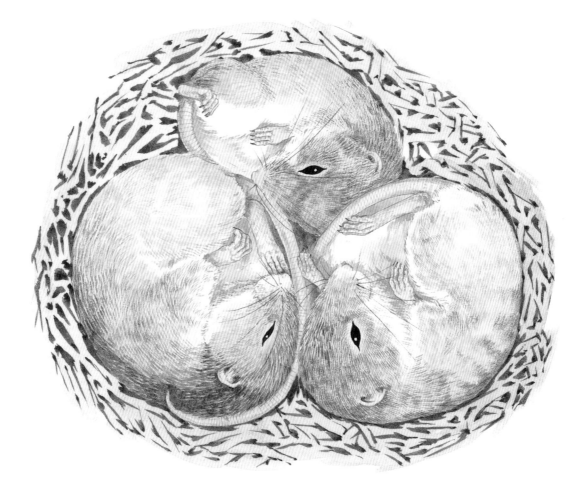

On the move

In the winter, some harvest mice move into holes dug by wood mice and other small animals. Clever mice spend the winter in barns and build warm nests in stacks of loose straw. Harvest mice nesting in reeds growing in water move to a drier area.

SHARING A NEST

When harvest mice are born late in the year, they sometimes spend the winter in the nest together. They stay close together, curled up in a ball. Sometimes the mother also stays with them. The more, the warmer!

Changing colour

Harvest mouse youngsters change from their dull grey-brown coat into a golden adult coat about a month after birth. The mouse grows thick new fur, or moults, in 35 days, but moulting can take more than 100 days.

HARVEST MOUSE CIRCLE OF LIFE

Harvest mice mate in the spring, and females usually have three litters of young every year.

While she is pregnant, the female builds a breeding nest on grass or reed stems.

Harvest mice are fully grown when they are about 45 days old.

26

After giving birth, mothers suckle their young for about two weeks.

Young mice explore outside the nest and can feed themselves when they are two weeks old.

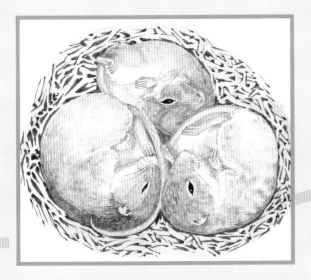

A few days later, the mother moves out to build a new nest for the next litter. The young mice stay in the old nest for a while.

27

Glossary

breed To mate and have young.

breeding nest A nest where an animal cares for its young as soon as they are born.

combine harvester A large farm machine that cuts and processes grain as it is driven through a field.

flexible Bendable.

groom To clean the fur or skin.

litter A number of baby animals born at one time.

mate When a male and female animal come together to make young.

moult To shed hair and make room for new fur growth.

nipple The teat of a female animal, from which milk can be sucked.

nursery A place where young animals are cared for.

pregnant With young developing inside the body.

prehensile Able to grasp things.

rodent A small mammal that gnaws food with its strong front teeth.

scruff The back of an animal's neck.

suckle To feed milk to a young animal.

territory The area that an animal defends against animals of the same kind, to keep them away.

thistledown Light, fluffy hairs that are attached to the seeds of thistle plants.

INDEX

An Appleseed Editions book

First published in 2005 by Franklin Watts
96 Leonard Street, London, EC2A 4XD

Franklin Watts Australia
45-51 Huntley Street, Alexandria, NSW 2015

© 2005 Appleseed Editions

Created by Appleseed Editions Ltd,
Well House, Friars Hill, Guestling, East Sussex, TN35 4ET

Designed by Helen James
Illustrated by Graham Rosewarne

Photographs by Corbis (David Aubrey, Tom Bean, Sally A. Morgan,
Ken Wilson; Papilio, Patrick Ward)

ISBN 0 7496 5840 1

A CIP catalogue for this book is available from the British Library.

Printed and bound in Thailand